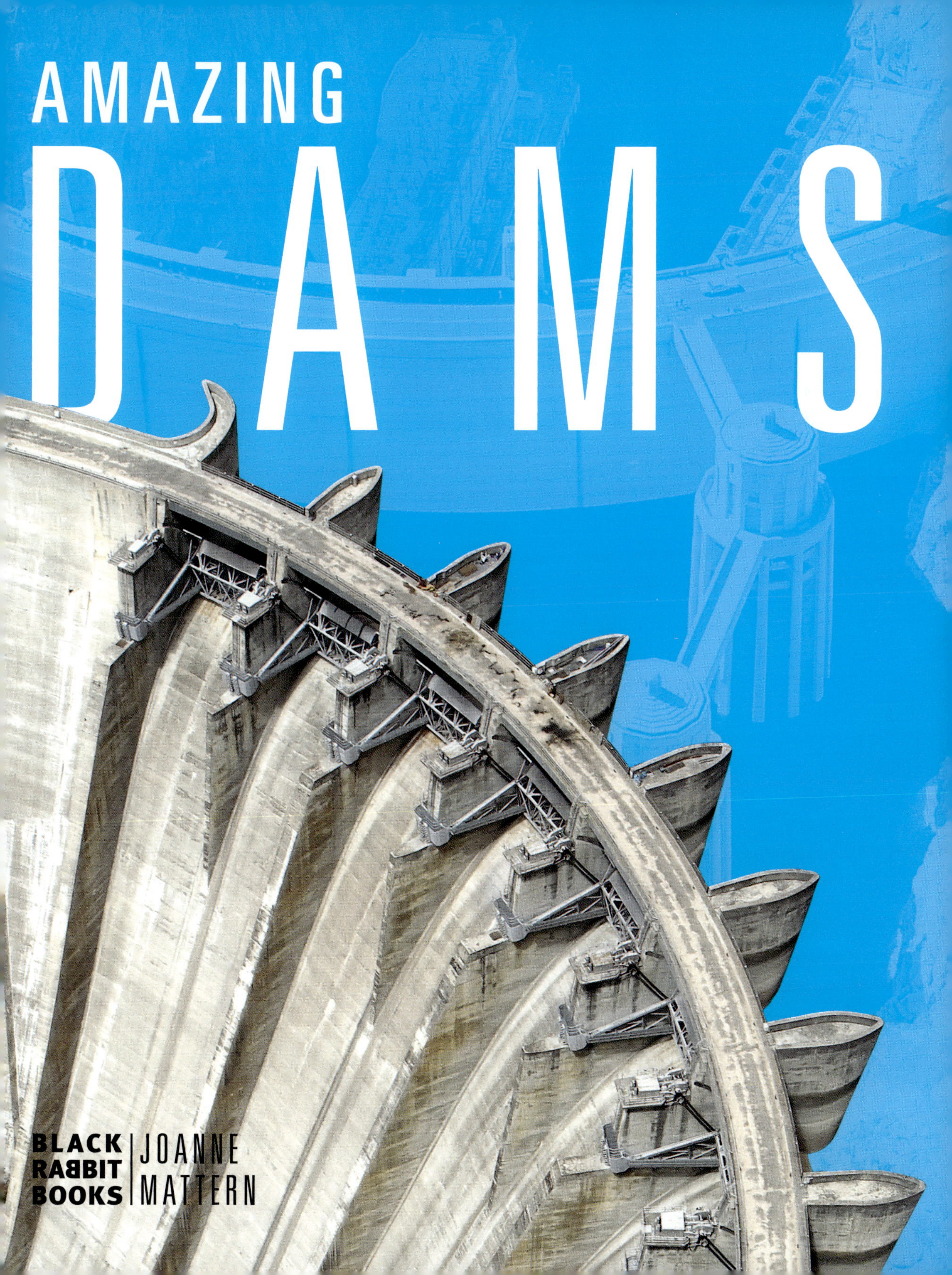
AMAZING
DAMS
BLACK RABBIT BOOKS
JOANNE MATTERN

TABLE OF CONTENTS

1

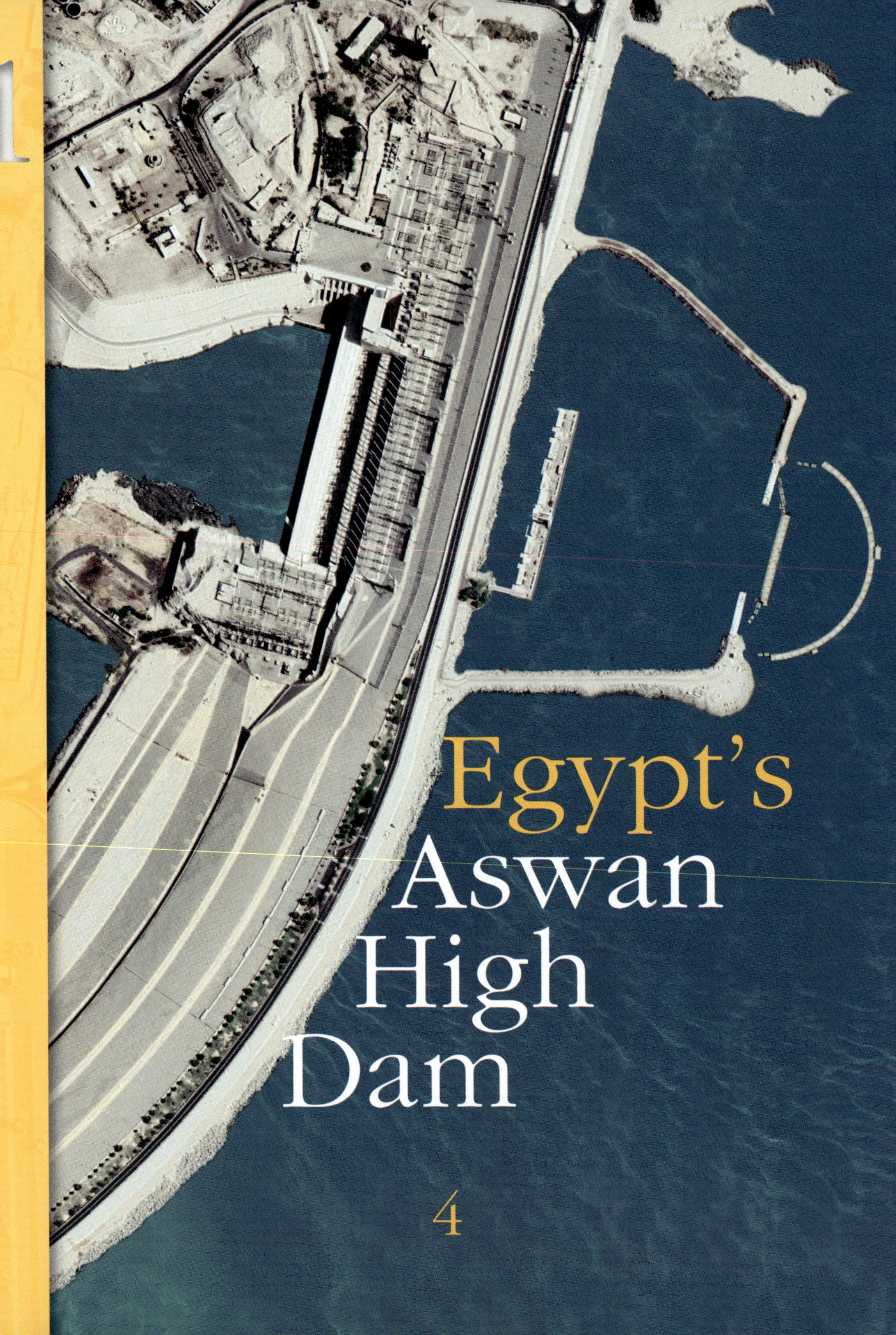

Egypt's Aswan High Dam

The Nile River is the longest in the world. People wanted to use it for power. In 1960, work began on a dam. It is the Aswan High Dam. It took 10 years to build. The dam is in southern Egypt. It cost about $1 billion.

The dam provides water for farming and drinking. It makes electricity. When it was built, many people were displaced. Important structures were moved out of the way.

Think About It

Is it right to make people move to build a dam? Why or why not?

2

Nevada's Hoover Dam

The Hoover Dam is in southeast Nevada in the United States. It is on the Colorado River. Construction began in 1930. It took six years to build. It was the tallest dam in the world. It is just over 726 feet (221 meters) high. Today, it is the second tallest in the country. Building the dam was dangerous. Workers blasted rocks with dynamite. They poured concrete into huge columns. The dam helps power cities in three states.

Did You Know?

The Hoover Dam created Lake Mead. It is the country's largest **reservoir**.

3

Switzerland's Luzzone Dam

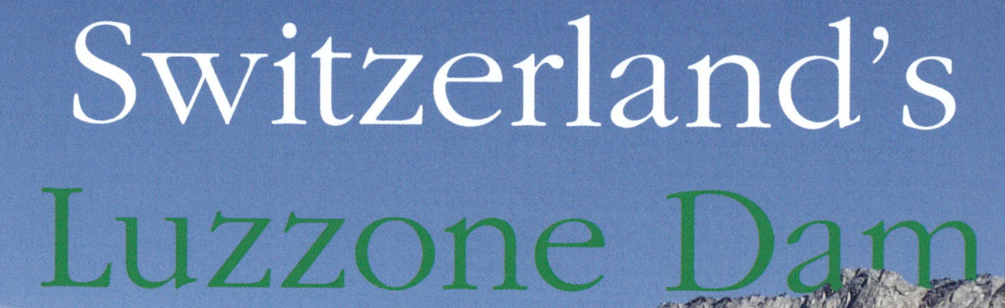

Dams control water. They also provide power. They can also be places to have fun. The Luzzone Dam is in central Switzerland. It has a climbing wall. The wall is 541 feet (165 m) high. It is the world's tallest human-made climbing wall. There are over 650 handholds. The wall gets steeper near the top.

The dam was built in 1963. It created the Lago di Luzzone reservoir. Mountain peaks surround the area.

Did You Know? Climbers use a ladder to reach the first handhold.

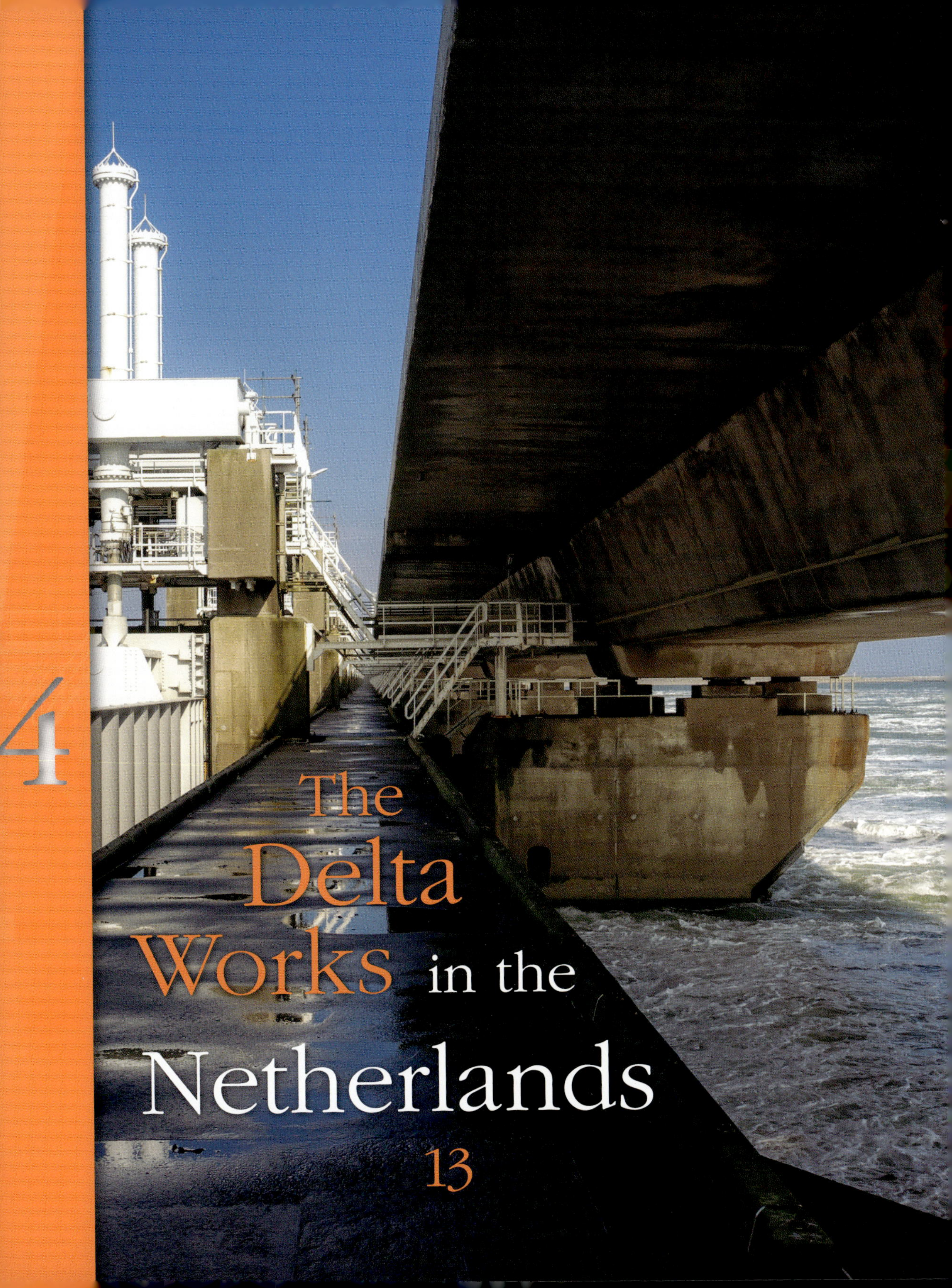
4
The Delta Works in the Netherlands
13

Much of the Netherlands is below sea level. Flood control is important. The Delta Works is a flood protection system. It's the largest of its kind.

The system is a chain of dams, **dikes**, and **levees**. There are also locks and floodgates. Surge **barriers** close during storms. They stop the sea from flooding inland. One of them is the Oosterschelde. It is over 5 miles (9 kilometers) long.

The Delta Works is important. It protects land for beaches. People can visit the natural areas.

Did You Know?

The Delta Works is one of the Seven Wonders of the Modern World.

Pakistan's Tarbela Dam

5

Most dams are built of only concrete. The Tarbela Dam is not. It is filled with rocks and dirt. The dam is in Pakistan. It was finished in 1984.

Tarbela is the largest earthfill and rockfill dam in the world. It rises almost 469 feet (143 m) high. It stretches 9,000 feet (2,743 m) across the Indus River. The dam allows the river to water crops. It also creates power.

Did You Know? Pakistan wants to make another dam. But many people live near the river. They are afraid they will lose their homes.

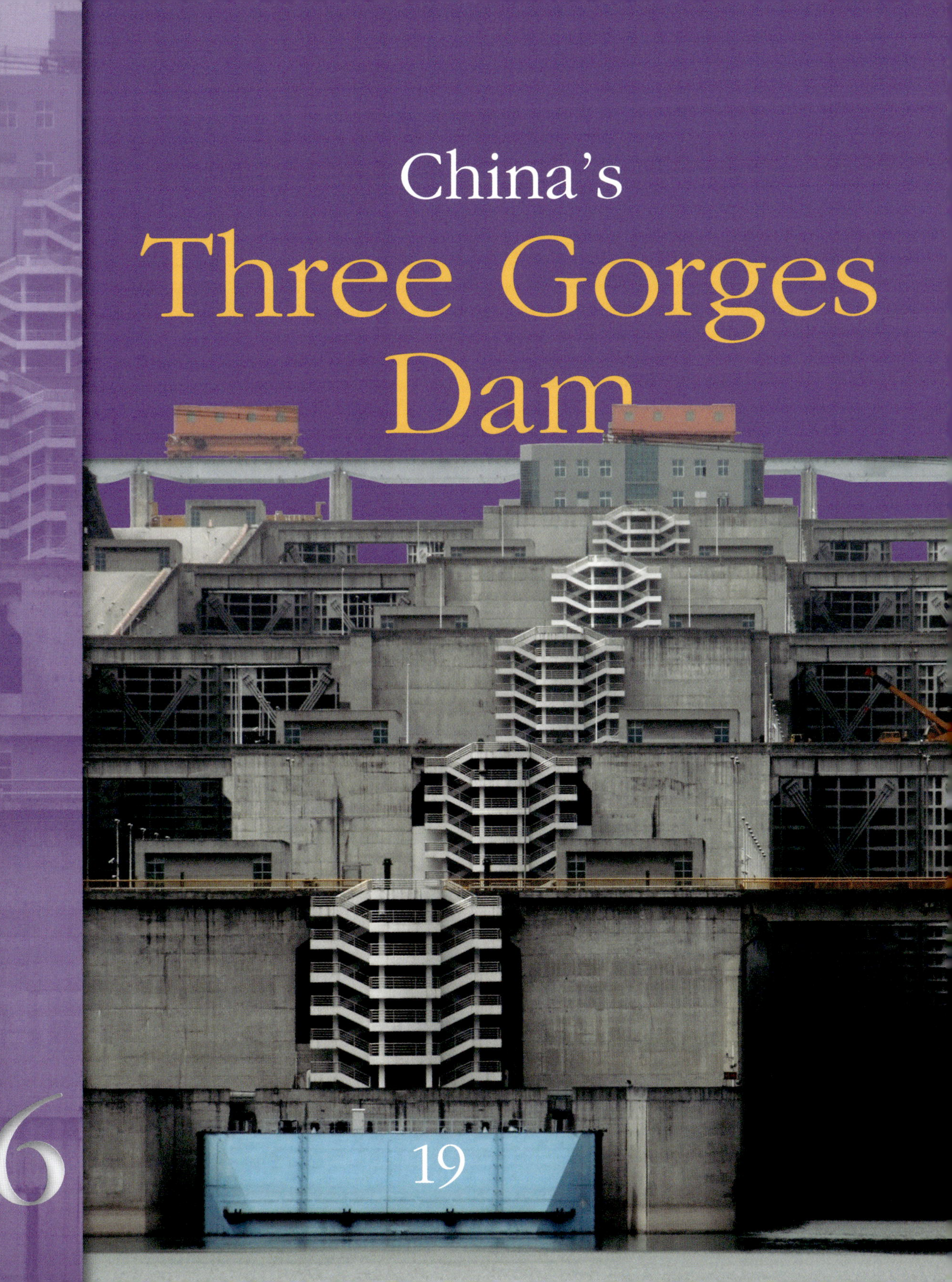
China's
Three Gorges
Dam
6
19

In 1994, China began building a huge dam. It crosses the Yangtze (yahng-ZEE) River. Three Gorges Dam was finished in 2006. It is the largest **hydroelectric** dam in the world. It is 7,770 feet across (2,335 m).

The dam creates plenty of power. It has 34 generators. But it also created problems. Over a million people were displaced. People worry the dam harms the environment. It shrunk the Chinese river dolphins' habitat.

Think About It How can a dam harm the environment?

Chinese river dolphin

MORE TO EXPLORE

WHERE IN THE WORLD

NEVADA'S HOOVER DAM

NETHERLANDS' DELTA WORKS

SWITZERLAND'S LUZZONE DAM

EGYPT'S ASWAN HIGH DAM

PAKISTAN'S TARBELA DAM

CHINA'S THREE GORGES DAM

MORE TO EXPLORE

FANTASTIC FACTS

The **Aswan High Dam** has enough rock to build more than 17 pyramids.

Boulder City was created to house the 5,000 workers who built **Hoover Dam**.

Workers had to change the course of the Colorado River to build the **Hoover Dam**.

Climbers pay a fee to get onto the **Luzzone Dam**. The cost includes a rental ladder to reach the first handhold.

In 2005, a powerful earthquake struck the **Tarbela Dam**. But the dam held tight.

There are ancient sites along the Yangtze River. Many homes were washed away by the **Three Gorges Dam**.

MORE TO EXPLORE

COOL COMPARISONS

How do these amazing dams compare?

Hoover Dam
completed in 1936

Luzzone Dam
completed in 1963

Aswan High Dam
completed in 1970

Tarbela Dam
completed in 1984

Delta Works
completed in 1986

Three Gorges Dam
completed in 2006

MORE TO EXPLORE

RESOURCES

Glossary

barrier (BARE-ee-uhr) A structure that blocks passage.

dike (DAHYK) A bank or mound of earth that is built to control water and protect from flooding.

displaced (dis-PLAYST) Forced to leave the area or country where you live.

hydroelectric (hye-droh-i-LEK-trik) Using waterpower to create electricity.

levee (LEV-ee) A long wall of soil built along a waterway to prevent flooding.

reservoir (rez-uh-VAWR) A lake used as a water supply.

Read More

Dittmer, Lori. *Hoover Dam.* Mankato, MN: Creative Education/Creative Paperbacks, 2020.

Mihaly, Christy. *Energy from Water.* Lake Elmo, MN: Focus Readers, 2022.

Index

TOP RANK is published by Black Rabbit Book, P.O. Box 227, Mankato, MN 56002 •

• Top Rank is an imprint of Black Rabbit Books. • Edited by Alissa Thielges • Designed by Danny Nanos • Photographs © Alamy: Christine Osborne, 18, imageBROKER/Manfred Stutz, 11, Patty Tse, 20; Getty: aaaaimages, 8–9, Lollo Riva, 12, Mor65, 2–3, 10, 21, 23, Philippe Turpin, 15, prill, 19, Zulfiqar Ali, 16; Shutterstock: burakyalcin, 4, 21, Dmitry Rukhlenko, 5, first vector trend, 17, Photomarine, cover, SAHAS2015, 14, seeyah panwan, 21, Timothy OLeary, cover, 6–7, 21, 23, Wut_Moppie, 13, 21, 23 • Printed in the United States of America

Library of Congress Cataloging-in-Publication Data: Names: Mattern, Joanne, 1963- author. |Title: Amazing dams / by Joanne Mattern. | Description: Mankato, MN: Top Rank, an imprint of Black Rabbit Books, [2025] | Series: Design marvels | Audience: Grades 4–6 | Identifiers: LCCN 2023058226 | ISBN 9781632357892 (library binding) | ISBN 9781645820680 (ebook) | Subjects: LCSH: Dams—Design and construction—Juvenile literature. | Classification: LCC TC541 .M38 2025 (print) | LCC TC541 (ebook) | DDC 627/.8—dc23/eng/20240130